Speaking in Public About Science

T0212432

Speaking in Public About Science

A Quick Guide for the Preparation of Good Lectures, Seminars, and Scientific Presentations

Ulysses Paulino Albuquerque

 Springer

Ulysses Paulino Albuquerque
Departamento de Biologia
Universidade Federal Rural de
    Pernambuco
Recife, Pernambuco, Brazil

ISBN 978-3-319-06516-8   ISBN 978-3-319-06517-5 (eBook)
DOI 10.1007/978-3-319-06517-5
Springer Cham Heidelberg New York Dordrecht London

Library of Congress Control Number: 2014939950

Printed on acid-free paper

Springer is part of Springer Science+Business Media (www.springer.com)

Preface and Acknowledgements

I became interested in public speaking, particularly as it concerns the communication of scientific facts, early in the course of my academic training. The reasons for that interest were quite diverse, ranging from the suffering exhibited by my colleagues (and me) upon presenting seminars and papers at the university to the disappointment that followed lectures delivered by renowned scientists who seemed to suffer as much as any beginner. I shared my ideas with Dr. Antonio Souto, a specialist in ethology and primatology, whom I profoundly admire to this day. Although he agreed with me on several points, he commented that perhaps the main problem was not with public speaking but with scientific writing. Twenty years went by, and now, I think that both of us were right.

I believe that writing and speaking in public are two significant skills that anyone aspiring to an academic career should learn and develop. The reason is that science is a universal enterprise, and as such, it is based on communication as the main strategy to structure and establish itself. As a consequence, and taking my own shortcomings into account, I decided to meet the challenge and write on this subject. First, I wrote a basic text as support material for a short course I taught my colleagues at a master program in plant biology. That text was the result of a challenge posed by my friends and colleagues, the biologists Delmira Costa Silva, Gardene Maria de Sousa, and Valdeline Atanázio da Silva, which led me to put the ideas occupying my mind into writing. While still a master's student, I sought to refine that first version and offered to teach a course on Basic Scientific Public Speaking to broader audiences.

I have been teaching basic courses regarding the public communication of scientific facts (scientific public speaking) on a steady basis ever since. The present book summarizes the content of those courses and is an augmented translation of the third Portuguese edition published in Brazil. The present version in English follows the same basic plan I devised when I was an undergraduate student to produce a light, quick, and practical introduction to the intended subject. I honestly expect, dear reader, that this booklet will help you face the challenges posed by academic public speaking. My intention was to compile information in a concise and objective manner to help anyone who needs to speak correctly and coherently about science as well as to apply such information to his or her professional practice. After 20 years of teaching courses on public speaking, I have met many different types of professionals and students seeking to improve their speech skills. The present book was revised and slightly augmented to include suggestions made by students and colleagues. Upon the request of my editor at Springer, I added a few notes regarding online resources that might be useful for public speaking.

A debt of gratitude is owed to many people who helped make this book possible. Thanks are due to Eric Stannard, editor at Springer, for supporting the proposal. I am especially grateful to the National Council for Scientific and Technological Development (CNPq) for financial support in the form of scholarships for scientific productivity given to me.

Ulysses Paulino Albuquerque

Contents

1 Introduction: Do We Need to Speak About Science? 1

2 How Are Your Communication Skills? 7

3 Efficient Oral Communication 11

4 Types of Scientific Presentations 15

    Seminars 16

    Conferences and Lectures 16

    Roundtables 16

    Oral Communications 17

    Bulletin Boards or Posters 17

    Dissertation Defenses 18

        Use Your Supervisor and Friends as Test Subjects 18

        Do Not Talk to the Wall 19

        Do Not Stand Still 19

        Attention to Posture and Vocabulary 19

        How to Avoid Awkward Situations 19

        Speak Fluently 20

        Amend Your Mistakes Without Stressing Them 20

5 Reflecting on Expository Practice 23

    Positive and Negative Predisposing Factors 25

    What? 25

How? 26

To Whom? 28

    Read Speeches 29

    Available Time and Speech Preparation 30

6   Scientific Method and Logical Construction
of Discourse 31

Explanation 32

Clarity 32

Completeness 33

Impartiality 33

Order 33

Accuracy 33

Objectivity 33

Simplicity 34

    Excitement and Enthusiasm 35

    Sincerity 35

    Persuasion 36

7   Composition of Scientific Discourse 37

Structure of Scientific Discourse 38

    Introduction 39

    Body 39

    Conclusion 40

Script Preparation 41

8   Basic Rules for the Use of Visual Resources 43

Rules to Observe 44

Use of Slides 44

Use of Blackboards or Equivalent Resources 45

9 The Scientist and His/Her Public 47

10 Some Exercises and Online Resources 49

Exercise 1. Training the Organization and Preparation
of Speeches 50

Exercise 2. Identifying Weak and Strong Points 51

Exercise 3. Organizing an Informational Discourse 51

Exercise 4. Organizing a Persuasive Speech 51

Exercise 5. Understanding Different Audiences,
Different Speeches 51

Exercise 6. Preparing the Introduction 51

Exercise 7. Developing the Speech Body 51

Exercise 8. Concluding a Speech 51

Exercise 9. Improving Your Oral Performance 52

Exercise 10. Preparing Visual Resources 52

Exercise 11. Re-reading a Story 52

Exercise 12. Exploring Lasswell's Model 52

Exercise 13. Using the Scientific Model 52

Online Resources 53

11 Supporting Materials 55

Text 1: I Am a Doctor, and You? 56

Text 2: What Is a Thesis? 58

What Is a Thesis? 59

What Type of Professionals Do We Want to Train? 60

What Do We Want for the Future?.................. 61

Text 3: The Science I Imagine and the Science There Is...... 61

References...................................... 65

Index... 67

Chapter 1

Introduction: Do We Need to Speak About Science?

U.P. Albuquerque, *Speaking in Public About Science*,
DOI 10.1007/978-3-319-06517-5_1,
© Springer International Publishing Switzerland 2015

In his book, "*Scientists must write*," Robert Barrass (1991) highlighted that the act of writing, as a part of day-to-day work, allows people to express their thoughts and then added that people with poor writing skills are at a disadvantage in learning as well as in their personal relationships. This idea is not novel. Nevertheless, that fact is neglected due to the little relevance attributed to formal instruction in the art of writing by our universities. For that reason, it is not reckless to assert, together with P. B. Medawar (1982), that "*most scientists do not know how to write*." The undeniable relevance of writing notwithstanding, many people are unable to express their ideas clearly or to support their arguments with the knowledge that they have and that was acquired through experience and learning.

The matters are no different regarding oral expression, but they are much more serious, to the point that we might restate the sentence above as follows: "*most scientists do not know how to speak*." Despite fully mastering a given subject, many people make a true cul-de-sac out of their speech. These people transform stimulating and interesting subjects into lackluster and demotivating presentations due to a lack of presentation attractiveness and appropriate planning. The art of public speaking in science is equivalent to the art of writing about science: it cannot be ignored or underestimated. Words, when properly spoken, have the power to attract the attention of, convince, entertain, and persuade people.

All these facts make the significance of oral communication for people active in several sectors of society clear. Appropriate communication that complies with the purity of logical and coherent speech is crucial. Writing is a relevant stage of scientific research because it precedes the definitive divulgation of the research results, and the oral communication of those results is not less relevant.

Oral presentations are a part of the everyday life of people involved with scientific discourse because such presentations are intrinsic to lectures, communications in scientific meetings, teaching, and so forth. University students are well aware of that valuable instrument as a consequence of the continuous requirements to present seminars and papers, which ought to be done successfully because they are a component of the assessment of performance in a given class. In reality, many skilled professionals are unable to hide their fear of facing an audience and expounding upon their ideas and research because they believe themselves unable to speak fluently on a subject even when they fully master it. Although speaking fluently does not merely consist of placing oneself in front of a group of people and saying something, that act is precisely what many people do. Thus, while these people believe they are explaining something, they are actually confounding their audience; while they believe they are motivating, they are demotivating; while they believe they are teaching, they are merely speaking.

Speaking correctly about science demands steady improvement of one's **vocabulary** by reading enriching works, of one's **posture** facing a certain group of people, and of one's **theoretical and practical knowledge** of the principles that characterize speaking about science. As any other task, speaking about science demands active initiative and a posture grounded on an uninterrupted quest for knowledge and improvement as a professional and a human being. In this regard, it is worth reading a fragment from the interesting autobiography of François Jacob (1988), a Nobel laureate in Medicine:

> I carry within, a kind of inner statue, a statue sculpted since childhood that gives my life a continuity and is the most intimate part of me, the hardest kernel of my character. I have been shaping this statue all my life. I have been constantly retouching, polishing, refining it. Here, the chisel and the gouge are made of encounters and interactions; of discordant rhythms; [...] all the sensations and constraints, marks left by some people and by others, by the reality of life and by the dream.[1]

The need to improve in scientific activity is too obvious to have to stress it here. The same is the case for oral presentations. Many people know that they cannot express themselves appropriately in public, and this limitation is the source of much frustration. This fact is illustrated by those barely passable dissertation defenses in which the candidate loses his or her voice from the mere awareness of being the focus of attention or illustrated by the wearying lectures in which the speaker abuses ambiguous verbal constructions, poorly argued ideas, and inadequate expressions in addition to common problems with elocution and gestural expression. Despite these problems, **very few people seek to improve their oral expression**.

When the abovementioned factors are not the problem, there are still problems related to the inappropriate use of visual resources, which otherwise are powerful aids for communication. When I was an undergraduate student, my class had a chance to perform a funny and critical play targeting the so-called OVERHEAD PROFESSOR. That character was a professor

---

[1] Translated from the Portuguese edition "La Statue Intérieure" published by Éditions Odile Jacob.

who would come in front of his or her class and perform a very dynamic reading of transparency sheets on an overhead projector having nothing to add to the sheets. Although this problem is a quite relevant matter, a thorough analysis of it is much beyond the scope of this book.

Young researchers become aware quite early of their need to improve as professionals upon learning to speak an intelligible and coherent scientific language and realizing its relevance for their professional careers. Together with the growth of that awareness, these researchers realize that their results have to be disseminated in writing and speech sooner or later, be they original or not. That point is addressed by P.B. Medawar (1982) in his advice to young scientists:

> It is psychologically most important to get results, even if they are not original. Getting results, even by repeating another's work, brings with it a great accession of self-confidence; the young scientist feels himself one of the club at last, can chip in at seminars and at scientific meetings with "My own experience was..." or "I got exactly the same results" or "I'd be inclined to agree that for that particular purpose medium 94 is definitely better than 93," and then can sit down again, tremulous, but secretly exultant.[2]

As a general rule, *the more exultant he/she will be, the better he/she presents his/her ideas.* Many skilled researchers and students are unable to make successful presentations but annoy themselves and their audience when they try to convey their research results using words. They suffer by merely thinking that they will have to speak in public, using as many excuses as possible to refuse or cancel presentations. Once, a colleague invited me to attend the presentation of a study that had afforded its author an award for its technical-scientific contribution, a fact of which we were not aware at the time. At the end of his talk, groups of people commented on the lack of interest of that study and on how painful it had been to listen to that speaker. Only much later, we learned of the award from a friend, who was an expert in that subject. Conclusion: the written version of the study was extremely interesting, but in its oral version, it failed to persuade anyone.

I have noticed that the actual problem quite often is not a lack of technical-scientific training to speak in public but carelessness and a lack of commitment to the audience. One should bear in mind that the public

---

[2] Translated from the Portuguese edition.

wants to understand one's work and to follow one's speech without this implying suffering for the speaker and for the audience. For that reason, spend appropriate time in preparing your presentations and in improving your speech.

Chapter 2

How Are Your Communication Skills?

U.P. Albuquerque, *Speaking in Public About Science*,
DOI 10.1007/978-3-319-06517-5_2,
© Springer International Publishing Switzerland 2015

Before moving forward, I would like to pose a challenge to you. Answer the questions below in full honesty, and then, assess your current state as a communicator in science. The questions apply to both expository practices and everyday communication in relationships with friends, colleagues, pupils, supervisees, supervisors, and so on. You might have some difficulty answering some questions, particularly the ones addressing experiences you have not yet undergone. In such cases, speculate on your possible behavior in the given situation.

	Yes	Sometimes	No
1. Do you usually expound your ideas in a timely manner, or do you always speak your mind without pondering on whether the context is appropriate?			
2. Do you probe your interlocutor (public, pupils, supervisors, among others) for his or her expectations before suggesting an idea or giving a speech?			
3. Do you start talks (chats, seminars, or lectures) by seeking to empathize with your interlocutor?			
4. Do you maintain a balance between speaking and listening (trying to talk less than you listen)?			
5. Are you concerned with confirming that your interlocutor understood your ideas properly?			
6. Do you prepare your presentations well ahead (or do you leave them to the last minute)?			
7. Do you usually plan your presentations (seminars, lectures, among others) thoroughly (or do you do the first thing that crosses your mind)?			
8. Do you usually think before speaking (or are you usually impulsive)?			
9. Do you indicate the presentation or communication aims before speaking?			
10. Do you carefully listen to opinions contrary to yours?			
11. Are you usually indulgent and polite when silly or unrelated questions are posed?			
12. Do you convey your ideas confidently and with thorough conviction?			
13. Do you try to stand in the listener's shoes when preparing a presentation?			
14. Do you always seek to add to the knowledge you have on the subject of the talk?			
15. Do you consult many references (personal or in the literature) when preparing a presentation?			
16. Do you argue without humiliating or attacking the other participant?			
17. Do you ponder long on your ideas before communicating them?			
18. Are you objective while speaking, and do you avoid irrelevant details?			

continued

continued

19. Do you usually treat people as your peers and respectfully?
20. When speaking, do you present as many new things as possible and avoid talking about what everybody already knows?
21. Do you organize your presentations by dividing them into main sections, i.e., introduction, body, and conclusion?
22. Do you try to imagine possible flaws and objections before making a presentation?
23. Do you always answer questions sincerely, eventually saying "I don't know" when applicable?
24. Upon preparing a presentation, do you try to match it to the target audience, i.e., do you adapt the vocabulary and content to your interlocutors?
25. Do you hold strictly to the intended subject, i.e., "talk about Brazil without travelling around the world"?
26. Do people who want to know your opinion on some subject constantly approach you?
27. Is it easy for you to express your ideas?
28. Are you normally understood by people (or do they ask you to repeat what you have said)?
29. Do you accept people who think differently than you?
30. Do you always look for people to cooperate and support one another?

Calculate your answer scores as follows:

Yes—3 points
Sometimes—2 points
No—1 point
  Add the scores together, and check the corresponding interpretation below:
30–49: You need to address this matter as soon as possible!
50–70: Fair. You must pay more attention to this subject
71–90: You are a good communicator. However, remember that you can always improve

Use this challenge to reflect on your attitude as a communicator. It might surprise you, but sometimes our problems with speaking in public are due not to technical-scientific flaws in the preparation of a presentation but to psychological and emotional factors. Quite often, our worldview and attitudes toward people are perceptible to the public. At the end of a lecture, e.g., you might have thought, "Gee, the subject was interesting, but the speaker was too arrogant." You might not be even aware of it, but sometimes your perception of a speaker as "arrogant" might affect your understanding of the lecture itself, i.e., build a "psychological" barrier against the discourse that was coming to you. For that reason, do not be hasty, but stop to ponder long, particularly on the questions you answered with a resounding "No"!

Chapter 3
Efficient Oral Communication

U.P. Albuquerque, *Speaking in Public About Science*,
DOI 10.1007/978-3-319-06517-5_3,
© Springer International Publishing Switzerland 2015

Oral communication is an efficient and quick way to communicate with our peers. Communication allows the establishment of affective ties, argument of points of view, and conveyance of ideas. In turn, those ideas might influence, conquer, inform, and persuade people. By communicating with other people in their field, professionals exchange experiences, disseminate the results of their research, and might elicit changes of opinion and consequently also of behavior.

For communication to be efficient and undistorted, it should be based on a relationship of reciprocity between the sender and receiver. Communication increases in efficacy and fruitfulness when the sender is polite and understanding of the interlocutor and supports the latter by means of encouraging words. Resistance is always raised to the reception of a message whenever the sender behaves as if he or she were superior to the receiver, behaves as if he or she were the owner of absolute truth, inhibits the receiver by using aggressive expressions or expression that shows that he or she is analyzing the interlocutor, or seeks to control the receiver's attitude. When these elements are present, the receiver—who is the target of the message—raises barriers and activates natural defense mechanisms that make him or her reject the message, regardless of its coherence.

In personal relationships, the elements mentioned above are even more relevant because the way one behaves with other people in daily life sooner or later also manifests in public communication in scientific discourse. Those elements determine the effectiveness of communication by influencing the interest in and proper reception of the message. Often, some people do not comply with the principles for good coexistence and communication, are unable to express their ideas clearly, or, even worse, consider others as undesirable for social coexistence. For that reason, you should question yourself regarding some features related to the improvement of communication and personal relationships:

Do I consider the people who I regularly interact with as equals?
Am I a good listener?
Do I use appropriate expressions when I address people?
Do I usually address people the way that I would like them to address me?

If most of the answers to those questions were "no," then something is wrong, and one should rethink one's attitude. By reflecting on these questions, you might discover the cause of your problems in oral expression and personal relationships.

When people fail to understand what you are trying to convey, the features below ought to be checked, as Krausz suggested (1986:41):

"Do I have a clear idea of what I intend to communicate?"
"Is my language appropriate to my interlocutor's linguistic universe?"
"After communication is over, do I check whether my interlocutor understood the message?"
"Is my tone of voice appropriate?"
"Do I select an appropriate timing and setting to communicate?"
"Am I concise, precise, and objective?"

When difficulties are present regarding the expression of ideas, Krausz adds the following features:

"Am I afraid of saying something wrong or irrelevant?
"Am I afraid of being mocked or rejected?"
"Do I believe in my own ideas?"

If you truly want to be a good communicator in any social area, you should pay attention to the abovementioned questions and answer them honestly. In addition to painting an accurate picture of your current status as a communicator, the answers to these questions facilitate the process of increasing self-knowledge.

> **The Ten Commandments of Good Communication (Krausz 1986:42)**
>
> 1. "Respect your interlocutor by being a good listener."
> 2. "Make your ideas clear before communicating."
> 3. "Define the actual goal of communication."
> 4. "Take the setting and people into account."
> 5. "Upon communicating, pay attention to the coherence between verbal and non-verbal language."
> 6. "Take the interests and knowledge of your interlocutor into account."
> 7. "Check and monitor the effects of communication."
> 8. "Try to understand and be understood (empathy)."
> 9. "Show interest in your interlocutor by keeping visual contact."
> 10. "Be concise, precise, and clear."

THINK ABOUT: by observing the commandments above, the efficiency of communication increases, and thus, these commandments should represent a constant source for meditation and careful consideration. By keeping those commands in mind, flaws in message transmission that originate in the sender might be diagnosed, such as "*lack of clarity, use of words maladapted to the given situation, predominance of secondary issues and neglect of the core, use of irony to mock the receiver, use of a categorical tone that diminishes the receiver*" (Silva 1991:32). Obstacles to message reception might alternatively originate in the receiver as a function of individual psychological or physical problems.

To conclude, analyze the obstacles to and positive features of good oral communication (Silva 1991:32):

Positive Features
Knowledge of the subject.
Clarity.
Presence (balanced bodily attitude).
Simplicity (gestural sobriety).
Pleasant voice and good diction.
Language adjusted to the audience.
Attention.
Self-control.
Confidence.
Availability.

Obstacles
Confusion of facts and ideas.
Systematic replacement of ideas.
Exaggerated gestures.
Monotonous or unpleasant voice.
Dispersion.
Stubbornness.
Mistrust.
Inability to listen to other people.
Partiality.

Chapter 4
Types of Scientific Presentations

U.P. Albuquerque, *Speaking in Public About Science*,
DOI 10.1007/978-3-319-06517-5_4,
© Springer International Publishing Switzerland 2015

There are several academic settings in which you might be called on to defend or present ideas or scientific findings.

Seminars

Seminars consist of the presentation of a given subject, which might be chosen by the presenter or by others. The aim of a seminar is to explore a given subject more deeply by approaching it from various perspectives. When seminars are conducted in a classroom, students read and present previously selected articles or topics consulted in textbooks. To perform well in seminars, the following pieces of advice are useful to keep in mind:

- Always do a little more than the prescribed reading on the seminar subject.
- Conduct your presentation in an interesting manner, and be wary not to include topics that will be addressed by other speakers.
- Reflect on the subject, build an opinion, and defend your opinion using arguments resulting from your reading and reflection.
- Do not extend over the allotted time to avoid interfering with the following presentations.

Conferences and Lectures

Both of these words are often used interchangeably to refer to presentations on a given subject with a definite time limit, which is usually recommended not to be over 60 min. Therefore, you must make the best use of the allotted time to expound your ideas. If you judge the allotted time excessive and that you will use less time, do communicate this fact to the organizers for the proper rearrangements to be made. Always avoid filling spare time with irrelevant topics. Once, I attended a lecture, the subject of which was quite interesting to me. The presenter devoted the first 15 min to speaking about how amazing his PhD program was, describing how he met many interesting people and showing some pictures of his travels, before addressing the proper lecture subject, which took only 15 min.

Roundtables

In roundtables, several individuals specialized in a given subject are reunited to discuss it from several perspectives or according to its variations. As a rule, two or three speakers are given 15–30 min to talk, and then, the public poses questions. Some particular features should be taken into consideration in regard to roundtables:

- Presentations should not deviate from the main topic.
- Presentations should strictly concern the indicated subject.
- Fruitless debates or contentions should be avoided, although this might be difficult to achieve when the topic is polemical. In addition, try to find out how the other speakers will address the subject to ensure that the presentations complement one another even when different points of views are argued.

When roundtables are not appropriately planned, quite often they tend to look like a patchwork of lectures addressing disconnected subjects.

Oral Communications

As a rule, symposia, meetings, gatherings, and congresses allot 10 min to the speaker's presentation followed by 5 min for the public to pose questions. Sessions of oral communications are included in scientific meetings for researchers and students to present the results of their theoretical or practical research. The analysis of the submissions for that modality of presentation takes the following items into account:

- Relevance.
- Theoretical and practical framework.
- Accuracy of the interpretation of results.
- The speaker's mastery of the subject.
- The study conclusions.

Focus your energy into conveying the core of your work. Always ask yourself: what is the impression (feelings or information) I want to make on people at the end of the 10 allotted minutes? Assess your presentation to remove everything that is nonessential for your message to be conveyed.

Bulletin Boards or Posters

In this modality, authors present their work as a full or subdivided poster according to the meeting rules, and that poster should include appropriate illustrations and text. Posters are exhibited in a predefined place chosen by the meeting organizers, and the authors should remain by their poster at appointed periods of time to talk with the people interested in the poster. Carefully elaborated posters are efficient tools to convey messages. In this regard, the following recommendations should be kept in mind:

- Posters should be graphically and esthetically presentable.
- The number of words should be limited to facilitate and motivate the reading of the text.
- The fonts should be uniform and without flourish.
- The background colors should be properly combined.
- The illustrations should be informational, and their quality must be high.

Posters should be elaborated to allow people to be able to read and understand the essential message by themselves. Doubts or additional information will be personally addressed by the poster author, who might take advantage of that occasion to stress the relevance or significance of his or her findings.

Dissertation Defenses

Dissertation defenses are a specific modality of scientific presentation that involves expounding on a research project that is variable regarding its content and level. Dissertation defenses are conducted at the end of undergraduate and graduate programs.

Although having different goals and degrees of comprehensiveness, essays and master's and doctoral dissertations or theses are publicly presented in the same manner. These works are studies that contribute to the development of science and afford their authors training in and mastery of the knowledge specific to a given field of science.

Because it addresses specific subjects, written work targets a small number of readers who are interested in a given theme. What about public presentations? Dissertation defenses are attended by a diversified public comprising experts and students, and for that reason, it is recommended to define the technical terms clearly when such terms cannot be replaced. Remember that the presentation does not target the dissertation committee alone but also all the people in the audience, who decided to attend because they have some interest, be it due to their personal relationship with the speaker or to their interest in the subject of the presentation itself.

Use Your Supervisor and Friends as Test Subjects

Whenever possible, rehearse your presentation in front of your supervisor or selected friends. With time and training, as well as their advice and comments, you will become more confident and master the subject better.

Do Not Talk to the Wall

Always talk to the audience, which includes formal (examiners) and informal (the general public) attendees. Quite often, speakers tend to talk to the wall while giving a slideshow presentation, and this attitude severs their ties with the audience, making the message seem uninteresting.

Do Not Stand Still

Speakers should move around, albeit moderately, while taking care not to stand in front of images projected. In small halls, it is advisable to stand at one side of the screen and to avoid turning your back on the audience while pointing to items on the slides. The use of laser pointers is advisable in larger halls because, in this case, the screen is usually quite far from the speaker.

Attention to Posture and Vocabulary

Because dissertation defenses are formal events, the language used requires particular attention. The language cannot be extravagant or vulgar; it should instead be simple. The inquiries made by the dissertation committee should be answered in a straight, objective, and clear manner.

Regarding the speaker's posture, to sprawl over the desk, to lean against the wall, or to sit in the lotus position is the utmost manifestation of inelegance. Speakers should sit elegantly, with the back straight, and politely write all the committee's suggestions and opinions in real time. In addition, it is recommended to answer or counterargue the examiners' assertions in a timely manner because to interrupt them while they are making a point is rude, even when their point of view is the opposite of yours.

How to Avoid Awkward Situations

The following should be your basic assumption: if your work is good, it will be approved. You collected the data, used appropriate methods, and accurately interpreted the results, and your supervisor, who is an experienced researcher, gave his or her approval: what can you be afraid of?

Nevertheless, some examiners might be tactless and not ponder on their words, rudely pointing to flaws in the work and identifying possible paths to improve it using strong expressions at the limit of impoliteness. The present text is not the place to discuss the motives for such behavior. Such awkward situations are prevented when the candidate and his or her supervisor nominate for the dissertation committee people whose usual behavior in defenses can be anticipated and who are remarkable for their politeness, integrity, and professional competence. Thus, all parties involved avoid unnecessary suffering.

Speak Fluently

Quite often, dissertation presenters neglect the continuity between topics but simply divide the presentation into sections. For instance, he or she announces the heading "INTRODUCTION" and then expounds on its content, and the same procedure is employed for the remainder of the sections (materials and methods, results, discussion, and conclusion). In addition to being uncomfortable to the audience, that approach makes speech inelegant. Rather than settling for such abrupt divisions, some words can be used to bridge the gap between sections. For instance:

The aim of our study is to establish the influence of. . . ***For that purpose****, we employed such resources. . .* ***And thus****, we obtained the following results. . .*

Nothing can be more boring and disappointing than to listen to a presentation as if it were literally read, up to the point that the audience can perceive the placement of commas and periods. For that reason, never perform slideshow presentations to merely read them aloud!

Amend Your Mistakes Without Stressing Them

When some sentence is wrongly formulated and hinders the understanding of the idea it conveys, do not stress the mistake made, but repeat the sentence immediately after in its correct form. In this way, the attention of those who noticed the flaw will be diverted from it to avoid missing the right idea by fixating on the mistake.

Assess Yourself

When rehearsing, try to assess the items below because continuous improvement in the oral expression of scientific matters depends on a thorough analysis of these factors:

1. **Expression features**
   Body posture
   Vocabulary
   Gestures (nonverbal language)

2. **Technical features**
   Clarity and precision of the introduction and aims
   Content development
   Language adequacy
   Use and adequateness of visual resources

continued

3. **Psychological features**
   Self-motivation (keeping the interest)
   Self-discipline
   Self-control

Chapter 5

Reflecting on Expository Practice

U.P. Albuquerque, *Speaking in Public About Science*,
DOI 10.1007/978-3-319-06517-5_5,
© Springer International Publishing Switzerland 2015

The transformation gradually undergone by people once they decide to improve their oral expression skills results from a process of self-knowledge and self-improvement. Even the individuals who have reached the upper levels of the scientific stair admit they still have much to learn and develop, including a most desirable trait: **scientific modesty**. This virtue constantly reminds a researcher that the knowledge that he or she has is nothing but a minute fraction of all that can still be known.

Presenters should be aware that scientific presentations are not mere expositions of some idea or of the results of a study to a group of people but that their aim is to CONVEY A MESSAGE. That message must be conveyed in an enthusiastic, energetic, and confident manner, while the speaker should attest to his or her knowledge and mastery of the subject for the message to reach the receivers unhindered.

Some hindrances arise from lack of mastery of the topic being presented, lack of appropriate techniques and resources, lack of confidence, or SCIENTIFIC ARROGANCE. Those factors distort the message because they build psychological barriers to a proper understanding of the message. Arrogance represents a major hindrance within the scientific milieu because the speaker becomes unpleasant to the audience from the very beginning. This situation is much more common than it might seem and is mostly due to problems intrinsic to the presenter that, nonetheless, do not suffice to justify some types of behaviors.

Once, I attended a lecture, the subject of which was quite interesting and timely. The hall was packed. The speaker took his place and started the presentation saying, "*Due to managerial problems, the subject of the lecture was slightly changed, and thus, I will talk long on...*" That statement discouraged the audience, but because the audience was quite indulgent, it listened to the presentation, although its expectations had changed. The presentation was a thorough disaster. The lecture actually delivered looked like a chapter of the speaker's autobiography, with his research and work, his interns, and so forth, in addition to direct personal attacks against some of the people in the audience. Everyone was shocked, and those who could left the hall discretely to then comment on the inappropriate behavior of the speaker, who, despite his professional competence, lost much of his good image. The fact is, people attend lectures because they want to learn more about its subject, listen to the views of the speaker on that subject, or increase their knowledge. However, who can stand a lecture like the one described above?

When someone speaks in public, be it a classroom seminar, an oral presentation of scientific research, or a conference, he or she must establish empathy with the audience. That goal is achieved when he or she complies with some principles, to wit, to be sincere when speaking and to show respect for the public. Nothing can be worse than hearing a lecturer say, "*I will not discuss this topic in depth because its level of complexity is high.*" That statement is nothing but a major exhibition of unjustifiable prepotency because the audience might include experts in that topic or people intelligent enough to understand it or even question the speaker's view. That example also shows that whenever a speaker does not intend to underestimate his or her audience, he or she should be wary of the meaning that people might attribute to some sentences that are not well formulated or are ambiguous.

Positive and Negative Predisposing Factors

Before accepting an invitation to speak in public, present a study, or deliver a speech, presenters should systematically ask themselves whether they are able to expound confidently on the subject of interest and whether they have enough time available to prepare it to avoid having to excuse themselves by saying something like, "*I apologize, but I hadn't enough time to prepare for this presentation*" or "*I'm too busy with other matters and had no spare time to prepare this seminar well.*" This type of behavior elicits negative reactions and makes people lose their interest in the presentation because the presenter does not inspire either confidence or certainty.

Thus, we should address the reflection that should precede the preparation of any presentation and that should include some relevant elements: What to SAY? How? To whom?

What?

The answer to this question is intimately linked to the goal of the presentation, i.e., what one intends to convey to the public: (1) Research results? (2) New information on a given topic? (3) An explanation of some subject? Those questions make the extremely informational nature of scientific presentations explicit. One should systematically pay attention to that which is publicized, especially in the case of technical or overly specific matters that might be of interest to a very small group of experts. Such subjects are unfit topics when one has to address a heterogeneous audience. In any case, Aristotle's advice still holds true: "*Think as wise men do, but speak as the common people do.*"

The programs of scientific meetings always make room, in either poster or oral communication sessions, for the presentation of highly specific information or methodological or technical innovations, thus allowing presenters to expose their results to a public with more homogeneous interests. Poster sessions are quite advantageous in this regard because only the people truly interested in a poster's subject will attend it. However, those sessions might be quite tiresome because they require the presenters to stand by their posters for over 1–2 h repeating the same speech once and again. In oral communication sessions, speakers are allotted up to 15 min to present their work, but they do so to a hall full of people waiting to listen to him or her as well as to the other speakers.

How?

Naturalness, friendliness, and energy are some of the essential traits required for a speaker to establish empathy with his or her audience. A speaker's words should be vibrant and vivid and should convey confidence and mastery of the subject addressed. Confidence is achieved when the speaker is PROPERLY PREPARED. A person earns the right to speak on a given subject only when he or she has mastered it and has full knowledge of it through experience or continuous learning. Therefore, you will always have something to speak about and a message to convey because all people exhibit thematic interests. The transmission of even the most exciting subject is hindered when the presentation was not well prepared or you speak in a voice too low or too loud, pause too long, speak too fast, or behave too monotonously or too energetically. Boring or monotonous speakers do not awaken their audience's interest and instead enervate it to the point that the audience stops paying attention even out of politeness. Although this problem leads to parallel chats and sleepiness, the cause of the latter cannot be systematically attributed to a poorly performed presentation.

The style of presentation is peculiar to each individual as a function of his or her skills and available techniques, and thus, presentation styles should not be imitated because "*the quality of an imitation is always inferior to the original.*" The process of self-knowledge and self-improvement leads to continual refinement of presentations, and each person gradually develops and establishes his or her way of expounding subjects. In this regard, it is useful to train and perform the following tasks suggested by Câmara Jr. (1993:45):

1. "*Define what we will say and consolidate our knowledge on it through reflection and research.*"
2. "*Distribute the subject as we deem to be the most interesting, clear and impacting manner.*"

Nervousness, from which not even the most experienced presenters are free, might hinder a presentation when it is not adequately controlled. The nervousness of beginners is betrayed by gestures driven by not knowing what to do with the hands or where to look, so they put their hands in their pockets and look to a point far away from the audience's eyes. Thus, they suffer and make their audience suffer too, and as the latter focuses its attention on the speaker's anguish, the message content slips into the background.

Disorderly motions, fads, and language vices (e.g., "ain't it," "ya know," "like") should be suppressed because although speakers often do not notice them, they attract the attention of the audience more than the message content. Once, I attended a lecture that was both motivating and elucidating, but the speaker finished every line of reasoning with "ain't it?" One could hear some attempts at suppressing laughter in the hall, while other attendees with mathematical proclivity counted the number of times "ain't it?" was repeated. One of these attendees announced that the total number was 110!

In addition to language vices, other problems concern elocution, including voice inflection and articulation as well as gestures, which lend expressiveness to speech. Balanced gesturing is a very desirable trait, and in fact, it has paramount importance because while you are saying one thing, your body might be telling something much different. Pierre Weil and Roland Tompakow wrote a book titled "*The body speaks*," which elaborates on that subject in a funny way. The lack of bodily expression due to nervousness or language vices (and sometimes "tics") gives rise to so-called PSYCHOLOGICAL CRUTCHES that serve as tension-release tools: playing in a disorderly and purposeless fashion with an object (pens, pointers, papers), looking constantly at a watch, drumming fingers, and so on. Dale Carnegie (1994) in "The quick and easy way to effective speaking" noted four facts regarding the fear of speaking in public, which is the origin of such nervousness, that are worth bearing in mind:

1. *You are not the only one who is afraid to speak in public.*
2. *Some measure of nervousness is useful because it is the means we use to prepare for unusual challenges.*
3. *Many professional speakers (presenters) never overcome their nervousness when facing an audience.*
4. *The main cause of your fear is that you are not used to speaking in public.*

Most often, nervousness is a reliable indicator of RESPONSIBILITY because it shows that the speaker wants to please and be properly understood. Any speaker will attain that goal if he or she prepares his or her presentation in an appropriate manner, speaks in a natural way and willingly, exhibits mastery of the subject, is confident, and presents the subject in a way that can be understood by anyone. You, dear reader, should bear in mind that a speaker should always face the audience in a confident manner and establish eye contact with the attendees all throughout the presentation. Regarding what to do with the hands, as mentioned above, there is a widely recommended technique to solve, at least partially, that problem: put one hand on top of the other and both in front of the body, and discretely press one hand against the other—you will notice that your tension gradually fades away. In time, that resource will no longer be needed, while efficient and fluent gesticulation gradually develops.

You Are Not Alone!

Look at the results of a social inquiry (cited by Lucas 2003) in "The art of public speaking":

Main fear	Response percentage
Being in a party with strangers	74
Delivering a speech	**70**
Having to answer personal questions in public	65
Meeting the boy/girlfriend's parents	59
First day at a new job	59
Being a prank victim	56
Speaking with an authority figure	53
Job interview	46
Formal dinner	44
Blind date	42

To Whom?

Some elements of a previously planned speech might have to be changed as a function of the setting and nature of the particular event in which it is delivered, to adjust to the new situation. For that reason, it is advisable to arrive to the venue at least 10 min early to adapt the presentation to the actual conditions found. In all circumstances, a speaker's posture should be balanced and sober, and his or her language and clothing should be suitable to the audience.

Sometimes, the public in scientific meetings is rather heterogeneous, and thus, the vocabulary used should be simple and easy to understand without being vulgar. It is advisable to bear in mind that the audience at scientific events includes people with different levels of expertise, including undergraduate and graduate students, experts in the targeted area, and nonexperts. For that reason, it is recommended to avoid the use of specific technical and foreign words because replacing them with simpler words helps the audience follow the line of reasoning. However, when some technical term cannot be replaced without compromising some key notion, the meaning of that term should be immediately explained and, whenever possible, illustrated with examples and analogies to make complex ideas clear. Any notion is more easily understood when it can be visualized and objectivized using an analogy or comparison.

Therefore, speakers should let the audience's expectations guide the message that they convey. For that purpose, one can sharpen one's perceptual skills by observing people's behavior and manifestations of interest.

Read Speeches

There is nothing more tiresome and annoying than a speaker reading his or her presentation. Read speeches might be somewhat accepted by the public under very special circumstances only or when that is the routine of a given scientific association or area of knowledge. As a rule, however, such practice is quite inconvenient because its monotony increases parallel to the length of the speech, and the audience discomfort grows together with it. Instead of reading speeches, it is recommended for speakers to prepare scripts to GUIDE their presentation. Learning a speech by rote is even worse than reading it because in addition to making the presenter "a speaking automaton," this practice might cause embarrassing situations when unforeseen accidents occur, such as a blank of memory.

If a speaker is able to speak fluently without the support of notes, that is the best option because it makes a favorable impression and the speech makes an impact on the audience, as experience shows. Once, the Russian psychologist Lev Seminovitch Vygotsky (particularly influential in the psychology of learning) delivered a speech that made deep impact on the audience because he did not have to resort to notes and spoke fluently.

Improper use of "lecture notes" excuses speakers from careful planning and learning. Once, a lecturer was repeatedly asked about a point that was not clear, and repeatedly, he would merely read from his notes and then ask, "*Is it clear now?*" Of course, it was not!

Available Time and Speech Preparation

Speakers should comply with the allotted time. While finishing a presentation a few minutes earlier is not a major problem, overextending always has serious consequences. As a rule, meetings include several speakers, and whenever one of them overextends, the time available for the remainder of the speakers is shortened or the meeting length has to be increased. Upon expounding on any subject, one should try to highlight the most relevant points. It is better to prepare a small number of topics, explain them well, and make them clear because presentations including too many topics have negative impacts because the topics are superficially explained.

Chapter 6

Scientific Method and Logical Construction of Discourse

U.P. Albuquerque, *Speaking in Public About Science*,
DOI 10.1007/978-3-319-06517-5_6,
© Springer International Publishing Switzerland 2015

My proposal is to construct scientific discourse based on the logic underlying the scientific method itself. These ideas are expounded in the present chapter and also in the following one. Scientific presentations must fully agree with some principles. Against some opinions, scientific discourse does not need to be grave and solemn to be considered serious and intellectually valid, nor does it need to be extravagant and full of unintelligible words.

Scientific notions should be explained in an accessible language to be properly understood. To translate technical vocabulary into a language that takes into account the audience heterogeneity, as described in the previous chapter, is always useful. Barrass (1991) observed that scientific writing should agree with the requirements of the scientific method, to wit, explanation, clarity, completeness, impartiality, order, accuracy, objectivity, and simplicity. The same requirements doubtlessly apply to oral presentations.

Explanation

Speakers should systematically take into consideration the level of knowledge of the targeted audience, i.e., they must establish whether previous explanations of a given subject are needed for the message to be better elucidated and fully understood. As a rule, very specific subjects demand timely explanations at each step of their presentation. The fragment below is from an abstract sent to a scientific meeting:

> In the present abstract, we described a recombinant protein of *L. chagasi*, which represents the carboxy-terminal fraction of the thermal shock protein with 70 kD molecular weight, for canine serodiagnosis.

In this case, the author might highlight some parts for further explanation, e.g., recombinant protein: what is it? How does it function? Serodiagnosis: what is it used for? carboxy-terminal fraction: what is it? Why is it used?

The short time allotted to oral communications in scientific meetings does not allow for thorough explanations. For that reason, presenters should highlight the parts of the speech that they judge to be more relevant for understanding the topic addressed. For that purpose, speakers should put themselves in the place of the listeners. However, one should be careful not to "preach to the choir" because to discuss items that everybody knows in full detail is counterproductive and tiresome.

Clarity

Clarity is a crucial element in the description of the aims and the definition of the problem to be discussed. Speakers should make the presentation **aims clear** for the public to follow their line of reasoning up to its logical conclusion. However, although a speaker might enounce the presentation aims clearly, sometimes he or she is not as clear in the rest of his or her speech. After such a presentation, one often hears the audience comment, "He spoke, and spoke, and spoke... but actually said nothing!"

Completeness

Completeness involves the full explanation of an idea, whereby every line of argument should be followed through to a logical conclusion. For that purpose, each argument or idea should be demonstrated by the corresponding proofs when necessary.

Impartiality

In the course of a presentation, speakers should state clearly the assumptions underlying their argumentation. Thus, speakers avoid enmeshing themselves in unsuitable opinions or preconceived ideas as well as hiding facts that contradict their views. Any stated opinion should be grounded on serious studies on the same subject. Expressions such as "*I believe this might be the case, however, I have never delved into that matter*" should be avoided because the audience might take such an assumption for scientific "fact."

Order

Sentences should be constructed and ordered to give a sense of continuity to the listeners. A logical order, studied by the speaker, allows for the information to flow naturally, thus allowing the audience to follow it without intellectual suffering.

Accuracy

Accuracy consists of the careful and precise use of words and expressions. Particular attention should be paid to definitions, historical data, personalities, and the methods used in an experiment because those items should be ACCURATELY ADDRESSED.

Objectivity

Objectivity is a quality that speakers should develop. Verbosity should be sacrificed for objectivity, i.e., one should address the intended topics in a straightforward manner. Look at the example below:

> The method used in this experiment consisted of the manipulation of 15 wooden stakes type... collected from the upper half of young and green

branches of a guava tree sown at a coconut tree plantation in the district of. . . Next, we painted them green using paint X, reference Y, code Z, and used them to delimitate a one-hectare area using #15 string.

Such statement is neither objective not precise, but plain boring! Banal details that make no positive contribution or that interfere with the understanding of the message should be discarded. If someone is interested in the brand of paint that the speaker used, he or she will ask about it.

Simplicity

Speakers should use simple explanations that comply with the full range of evidences. That criterion is known as Ockham's razor. Ockham was an English philosopher and theologian from the fourteenth century whose principle was that one should avoid *unnecessary* multiplication of problems and accumulation of hypotheses.

Simplicity prevents the use of pompous words, bizarre terms, extravagant explanations, and verbal polishes that add nothing but instead make the ideas conveyed unclear.

Do Not Forget

When addressing the linguistic style to be used in a scientific presentation, some requirements should be taken into account:

Brevity or Objectivity

Speakers are concise or objective when they omit irrelevant details in their presentation and employ only the words needed to convey an idea. Topics should be addressed in a straightforward and simple manner. As Horatio said, "*Be brief and you will be pleasant.*"

Clarity

A presentation loses clarity when the speaker uses unnecessary and ambiguous words with variable interpretations, which confound the audience.

Accuracy

A speech is accurate when the words used express exactly the speaker's thought. Some expressions fare better when they are supported by numerical data, as, e.g.,:

continued

> *Most students from the biological science course support the inclusion of extracurricular activities in the official syllabus.*
>
> It is more precise to state:
>
> *Approximately 80 % of the students from the biological science course support the inclusion of extracurricular activities in the official syllabus.*
>
> **Consistency or Unity**
>
> The logical organization of the sequence of topics in a presentation should be clear. The links or mutual dependence between all the addressed topics should be evident because, in this way, the presentation will convey a sense of order that is extremely pleasant to the audience.

TO ALSO TAKE INTO CONSIDERATION—None of the abovementioned principles and procedures coherent with scientific thinking suffice to ensure that the quality of a presentation will be high or that the speaker's performance will be satisfactory. That information contributes to making the flow of ideas coherent and to awakening a favorable reception in the audience. However, other factors should also be taken into consideration.

Excitement and Enthusiasm

Words increase their strength and vitality when the speaker is enthusiastic about the work that he or she does and wants to talk about. We become easily enthusiastic when a research project is highly satisfactory and pleasing. However, this enthusiasm is not the case when we are "compelled" to talk about an uninteresting subject, as happens quite often when the subjects of seminars are distributed in undergraduate or graduate programs. Under such circumstances, the speaker should look for the topics most attractive to him or her. If this approach does not work, it is advisable to try and achieve a more thorough understanding of the subject and its relevance because the speaker will then naturally succeed in defending some point of view with full enthusiasm.

Sincerity

Together with honesty, sincerity is a much-appreciated quality in speakers because it affords a chance for the speaker to inspire trust in the audience, which will react favorably by paying attention. In fact, the best strategy to obtain the public's goodwill and respect is to be sincere. In contrast, when speakers behave as if they were superior and sin against **scientific modesty**, they awaken animosity in the audience. As a rule, such speakers are judged negatively despite their acknowledged competence. In most cases, that type of behavior shows the speaker's lack of self-confidence as well as significant internal conflict resulting from wearying experiences that were poorly assimilated or overcome.

Persuasion

The public's interest should be awakened by giving the subject its proper value. According to the circumstances, one should indicate the implications, relevance, possibilities, and, eventually, the theoretical values of the topic. When evidence is provided in a convincing manner and appropriately presented, the audience will trust the speaker's words.

Chapter 7

Composition of Scientific Discourse

U.P. Albuquerque, *Speaking in Public About Science*,
DOI 10.1007/978-3-319-06517-5_7,
© Springer International Publishing Switzerland 2015

Some procedures are recommended for the elaboration of scientific discourses:

1. **Subject delimitation and definition**: this step comprises a study of the main message that one wants to convey, i.e., the CORE IDEA or SUBJECT. No subject can be so broad as not to be subject to proper delimitation of its most relevant parts for the purpose of presentation within a definite period of time. In this step, the expected aims and the means needed to achieve them are defined.

2. **Subject exploration**: subjects should not be too broad or too superficial, but they must include the full scope of data needed to attain the previously established aims.

3. **Comprehensive study of the subject**: this component is achieved by consulting the literature or more experienced researchers. To have more knowledge on the subject of a speech is always desirable.

4. **Reflection on the subject**: namely, on the subject's relevance, applications, implications, and so forth. This step involves addressing the subject from different perspectives, paying attention to issues encompassing it, and comprehending the various points of view put forward.

5. **Subject mastery**: it is always advisable to address a subject that is familiar to the speaker because, in that case, he or she will speak with full mastery of the presentation topic.

The procedures described above apply to several types of scientific presentations. Some such presentations are characterized by "secondary divulgation" because the presentations are grounded on data provided by third parties; such is the case of some classroom seminars conducted during undergraduate or graduate programs. Other presentations aim at providing a more encompassing view on their subject, and thus, the author's data are intertwined with those from other researchers. Finally, in one further type of presentation, the presenter reports his or her original, eventually innovative data, such as technical or methodological novelties, personal discoveries, or a novel approach to a subject. The particular presentation type notwithstanding, the organization of presentations should be as careful as the writing of articles intended for publication.

Structure of Scientific Discourse
Any presentation is usually divided into three sections, which might require internal alterations as a function of the subject, presentation aims, and circumstances.

Introduction

The introduction forms approximately 5–10 % of a presentation. The aim for this section is to awaken the audience's interest and to inform the audience about the subject that will be discussed. The introduction should indicate the presentation subject and aims in a clear and simple manner to orient the audience and guide the speaker along the presentation.

> Without it, a presentation becomes disturbing because it meets an environment more or less still unadjusted. Even when the audience has a previous interest in the upcoming presentation and is well predisposed towards the speaker, the latter's presence and the novelty of the experience conspire to hinder the audience from focusing its attention immediately on the subject; this is to say, a perceptual delay is generated, the result of which is that a basic component of the presentation body is at best lost (Câmara Jr. 1993:47).

This step allows the audience to obtain an idea on the subject and connect with the speaker, who, in turn, will earn or not the public's attention and empathy as a function of the way he or she starts the presentation. For these reasons, the introduction should be:

1. Proportional to the remainder of the discourse
2. Clear, objective, and extremely well defined and prepared
3. Attractive
4. Carefully elaborated because "the first impression is the last impression"

Body

The body is the section in which the speaker makes the presentation itself by describing ideas, theories, arguments, assertions, and proofs. The body represents 85–90 % of the presentation and consists of the unfolding of the core idea. All the assertions made should come together with the corresponding arguments and proofs, when required, especially in the case of polemical or controverted subjects. The full content should be presented as connected and mutually dependent to preserve the unity of the discourse.

All the data and examples to be used should be carefully checked because a poorly chosen example might contradict the speaker's argumentation.

Under some circumstances, the presentation body must be subdivided into well-defined parts to answer some questions characteristic of scientific research:

Materials and Methods (Methodology): How? Where?
This part comprises a thorough, clear, but concise description of the means, equipment, and materials used in the study.

> A speaker can be a bore not only for being insufferably prosy or because his work is intrinsically dull but because he goes into quite unnecessary details into matters of technique. Sometimes, it is judicious to spare an audience the details. If it is important to know and if the audience wants to know the order in which the speaker dissolved the various ingredients of his nutritive culture medium, he will be asked immediately after the lecture or privately later on (Medawar 1982:62).[1]

Results: What Did the Study Find?
The results section contains a description of the findings, observations, and experiments.

Discussion: How Were the Results Interpreted?
The discussion contains objective comments on the results aiming at examining and discussing the data presented. Information about the results of other studies is added for the purpose of comparison.

Conclusion
This section is a highly relevant part of presentations, but it is quite often neglected. It is advisable to devote 5–10 % of a presentation to a conclusion section. Conclusions should make an impact and help the audience keep a coherent and long-lasting idea of the subject addressed because one tends to remember better what was said at the beginning and the end of a speech. As a rule, the conclusion is a summary of the main ideas. For that reason, the conclusions should be:

1. Objective
2. Expressive
3. Categorical

---

[1] Translated from the Portuguese edition.

Tips

1. Upon elaborating some types of scientific discourses, such as the ones devoted to the dissemination of study results, some basic questions should be formulated to structure the discourse in a coherent manner:

   What for? For whom?—**aims**.
   Why?—**justification**.
   How? When? What with? Where?—**methods**.
   What was found?—**results**.
   How was it interpreted?—**discussion**.
   How were findings and interpretation summarized?—**conclusion**.

   In this way, a discourse becomes intelligible and attractive to the audience, who becomes able to follow the discourse from introduction to conclusion without mental suffering.

2. **Think and write.** It is highly advisable to prepare a speech as if it were an article or an essay because writing helps to put order to the ideas before speaking in public. When time does not allow writing the full presentation, making a list of the topics that will be addressed has a similar effect. In this way, the speaker's fluency, spontaneity, and eloquence are improved. However, speeches should not be learned by rote because that practice has the opposite effect.

Script Preparation

Once a speech is ready, the next step involves the elaboration of scripts. Several types of scripts might be used to guide a speaker during a presentation, among which we find the synthetic and the expanded scripts to be particularly interesting. Synthetic scripts consist of lists of the topics that will be addressed in their proper order. That same basic design is also used in expanded scripts, the difference being that some items are more thoroughly described. In both of these cases, indicate in the script the moments when you will use a given resource, such as projecting an image or making a pause apparently meant to tell some fact related to the subject but whose true aim is to relax the public. The following is an example of an expanded script.

**Example of an Expanded Script
(for a 20–30 min presentation)**

SUBJECT: Relationships between insects and plants.

Introduction

When someone is first inquired about the relationships that insects might establish with plants, he or she will immediately be given a list of harmful interactions that also affect man. But, are the relationships between insects and plants always harmful for plants?

In the present time, insects represent the largest group of animals, their number being larger than that of any other type of terrestrial animal. Insects evolved in many directions to adapt to life in different habitats. In this seminar, we will see that such characteristics enabled insects to establish not only interactions considered harmful for plants but also many others that are beneficial to plants.

1. Harmful interactions
   (a) The insects' herbivorous habit (show image 1).
       Plants' defense against herbivory.
       Main orders of herbivorous insects.
   (b) Xylophagous insects and their main orders.
       Example: harlequin beetle (*Acrocinus longimanus*).
   (c) Insect vectors of plant diseases (show image 2).
2. Beneficial interactions
   (a) Cross-pollination and evolutionary success of angiosperms.
   (b) Symbiosis between insects and plants (show image 3).
       Example: symbiosis between ants and *Acacia* species (*pause to tell an experience related to the subject*).

Conclusion

As we saw, insect–plant interactions represent a long co-evolutionary process, which resulted in a complex mechanism of relationships, some of which harm plants, but some of which also benefit them. This intimate association has allowed, among other factors, for effective pollination, the success of flowering plants, and the appearance of the world as we know it today.

Chapter 8

Basic Rules for the Use of Visual Resources

U.P. Albuquerque, *Speaking in Public About Science*,
DOI 10.1007/978-3-319-06517-5_8,
© Springer International Publishing Switzerland 2015

The fact that people learn more efficiently what they can simultaneously see and hear and can immediately apply into practice has been well established. Therefore, visual resources, which are elements that help to stimulate people and to visualize an idea, have paramount importance. Some of the visual resources most commonly used in scientific presentations are projections (slides, for instance). Black- or whiteboards are also frequently used; however, as a function of the circumstances, writing on such boards during a presentation might be impractical and time-consuming.

Despite the acknowledged relevance of visual resources, their improper use might cause effects opposite to the intended ones. The advice below is restricted to the use of projections.

Rules to Observe
Select the visual resources as a function of the aims to be attained and the practicality of the resource according to the circumstances.
The function of projections is to facilitate the visualization of some idea or image; therefore, slides should not include long texts.
Determine the time available for the use of resources.
Acquaint yourself with the equipment before using it.

TO NOTICE: When the rules described above are carefully observed, the goals for the use of visual resources are fulfilled, to wit:

- Motivate and awaken the audience's interest.
- Visualize and concretize some of the presentation topics.
- Illustrate ideas and notions.
- Make the presentation more clear.

Use of Slides
Regarding the use of slides, Barrass (1991:177–178) recommends:

"Do not show too many slides."
"Do not show a slide if it includes too much detail or anything that is not relevant to your talk. Use one slide to convey one message and make the message brief, clear and simple so that it can be understood quickly."
"Do not show a table if it has too many numbers or if the numbers are so small that some people cannot read them."
"Arrange the slides in the same order as your notes."
"Make sure that each slide is the right way up and the right way round."

"Give the audience time to look at each slide and then help them to interpret the slide."

"Remove each slide as soon as it has served its purpose, so that it is not displayed while you are trying to interest your audience in something else."

To which one might add:

- Always explain that which is shown on the slide.
- The slides should contain information or illustrations that explain a message and help visualize objects and notions. Therefore, the inclusion of short texts, such as the word "introduction" at the start of a given topic, does not make sense.
- Make sure that the slides are legible and that their visual and esthetic quality is high. It is better not to show substandard slides.
- Do not replace speech with the exhibition of slides.
- Make the text uniform regarding font and size.
- Make the best use of the available space when distributing the text; always use few words.
- Take the audience into consideration: present text and figures in a size that is appropriate to be clearly seen at the back of the hall.

Golden Rules!

A slide is usually seen for less than 30 s, so its impact has to be immediate. For this reason, figures for slides must be specially simple and succinct. A good slide makes no more than three points, and these points augment. Emphasize, and explain the speaker's words.

Sometimes speakers try to make one slide do the work of many. The result is visual confusion. For complicated subject matter, it is far better to use two or three simple figures than one complex and usually cluttered and unclear figure. Your audience may fall asleep while you are spending minutes trying to clear the confusion.

Reproduced from Briscoe (1995:117).

Use of Blackboards or Equivalent Resources

When properly used, blackboards are extremely useful in scientific presentations that do not impose some limitations on speakers. However, as a rule, this resource tends to be very poorly used because its functions are not taken into consideration.

Functions of Blackboards and Equivalent Resources

- To show schemes, summaries, and illustrations.
- To record few data and brief information.
- To reduce the speech's level of abstraction.

When using this resource, speakers should pay attention to the following recommendations:

- Write large and legible letters.
- Use mostly white chalk and other colors only to emphasize or highlight significant points.
- Do not turn the full back to the audience. Place yourself in a way that the audience can see what is being written.
- Start writing at the top of the blackboard.
- Erase the blackboard after the thing written/drawn has fulfilled its goal.
- Erase the blackboard from the top downwards.

Chapter 9

The Scientist and His/Her Public

U.P. Albuquerque, *Speaking in Public About Science*,
DOI 10.1007/978-3-319-06517-5_9,
© Springer International Publishing Switzerland 2015

Upon facing an audience, a scientist trained in self-discipline and respectful of his or her public should keep a true cordial posture under usual circumstances. In that case, it is recommended:

· To be indulgent with questions that are too simple or unrelated to the presentation's subject.
· To repeat a poorly elucidated topic as many times as asked for.
· Not to promote unfruitful discussions or debates.
· To act naturally and pleasantly when realizing that a question was posed to test his or her knowledge.
· To provide a brief response to the questions posed by the audience.
· To admit not knowing the answer to a question rather than giving a false or evasive response.

To conclude, a piece of wise advice by Peter Medawar:

> Scientists should behave in lectures as they would like others to behave in theirs. It is an inductive law of nature that lecturers always see yawns and a fortiori those hugely cavernous yawns that presage the almost complete extinction of the psyche. The same goes for anything else that may distract a lecturer (which may, of course, be the intention): sibilant whispers, ostentatious consultations of watches, laughter in the wrong places, slow, grave shakings of the head, and the like (Medawar 1982).[1]

Keys for Success Behind the Podium

· Have a short and engaging title.
· Speak with a strong voice and precise diction.
· Share genuine enthusiasm for the subject.
· In the introduction, acknowledge the audience field of work and knowledge base.
· Teach the audience with sample description of the material.
· To remain in control of the presentation, exhibit minimal reliance on the images.
· Preview, highlight, and "spin" each image.
· Give future directions succinctly, with a specific example for each.
· Repeat each question from the audience and keep answers brief.
· End the session on (or under) time.

Reproduced and modified from Morgan and Whitener (2006:81).

---

[1] Translated from the Portuguese edition.

Chapter 10
Some Exercises and Online Resources

U.P. Albuquerque, *Speaking in Public About Science*,
DOI 10.1007/978-3-319-06517-5_10,
© Springer International Publishing Switzerland 2015

Exercise 1. Training the Organization and Preparation of Speeches

In the selected text, try to identify: (1) The core idea. (2) The main points discussed by the author. (3) The point of view that grounds the author's ideas. As an aid, first read the text below on discourse plans and points of view. Then, choose any of the texts in Chap. 11.

Discourse Plans and Points of View

The planning of any discourse depends on the point of view adopted by its author. See Vanoye's (1993:170) considerations as applied to the subject of "bullfighting":

(a) *Historical point of view.* The plan should follow a chronological order; the presentation will discuss the origins of bullfighting and its function; it will analyze the current features of bullfighting in association with the past history, traditions, folklore, and various influences.

(b) *Technical point of view.* The various elements of bullfighting should be reviewed. Although a gradual plan is difficult to design, one might, e.g., start from the animal (breeding and preparation of bulls), then shift to man (bullfighting school), and move from the technique itself to the art of bullfighting.

(c) *Sociological point of view.* Starting from the required historical elements, the presentation will analyze bullfighting in its social role. The topic might be addressed from the bullfighters' point of view and also from the public's point of view (bullfighting as catharsis; the bullfighter profession as a factor of social advancement, similarly to the situation of professional soccer players in Brazil).

(d) *Literary point of view.* Bullfighting should be addressed as a literary motive, a kind of myth developed in legends, novels (e.g., Blasco, Ibañez, Ernest Hemingway), and movies.

Based on the same identified subject, choose one of the texts in Chap. 11 of this book, prepare a speech from a different point of view, and present it to friends or work colleagues.

Exercise 2. Identifying Weak and Strong Points

Make a list of a speaker's positive qualities and flaws. Next, critically assess your strong and weak points. Establish goals to improve the features that you consider to be undesirable.

Exercise 3. Organizing an Informational Discourse

Select one of the following subjects, or use the Support Materials, to prepare an informational speech: violence, culinary arts, fashion, the environment, or public health. Next, focus on delimitating your approach by answering the following questions: what is the overall aim of your speech? What are its specific goals? What is the core idea that will be addressed?

Exercise 4. Organizing a Persuasive Speech

Based on the instructions given in the exercise above, prepare a persuasive speech on the same subject. Then, compare the aims and core idea of both of the speeches. Remember that because the purposes of the speeches are different, their aims differ.

Exercise 5. Understanding Different Audiences, Different Speeches

Plan two speeches according to the criteria indicated below:

Subject: Adolescent pregnancy.
Audience 1: adolescents.
Audience 2: parents of adolescents.
Audience 3: adolescents (80 % female, 20 % male).
Audience 4: adolescents (20 % female, 80 % male).

Exercise 6. Preparing the Introduction

For the subject selected for exercise 3, prepare a draft of a captivating and objective introduction that explains the aims of the speech.

Exercise 7. Developing the Speech Body

For the subject used in exercise 6, prepare a draft of the speech body, taking all the elements discussed in this book into consideration.

Exercise 8. Concluding a Speech

For the subject used in exercise 6, prepare a draft of a categorical conclusion that reinforces the idea that you aimed to convey.

Exercise 9. Improving Your Oral Performance

Select a text, preferentially poetry, and read it aloud. Try to emphasize its most striking parts. Read slowly, carefully analyzing the parts. Express feeling in the most interesting parts.

Exercise 10. Preparing Visual Resources

Select any of the exercise speeches to plan the use of visual resources. Do not be hasty. Thoroughly analyze your speech to identify the points to which visual resources might make a relevant contribution. After making the presentation, assess the results. Ask for your colleagues' opinion. You will notice that some of the resources used might be replaced, improved, or eventually discarded.

Exercise 11. Re-reading a Story

In a group or alone, read a story slowly and aloud. Next, try to reconstruct the story without consulting the written text. This activity contributes to the development of the memory and synthesis skills.

Exercise 12. Exploring Lasswell's Model

Select a subject, topic, or event to prepare a short presentation. Lasswell's model is a powerful tool to improve the objectivity and clarity of ideas. Apply the model below to the construction of your speech:

- Who?
- Says what?
- To whom?
- Why?
- Where?
- When?
- How?
- With what results?

Exercise 13. Using the Scientific Model

Follow the same instructions as in exercise 12 but including the following items:

- Elements in the situation.
- Explanatory hypothesis.
- Hypothesis verification (proof and counterproof).
- Solution.

Online Resources

Although I believe that scientific discourse is structurally different from other varieties of speech, some basic elements for satisfactory communication are common to all subjects and modes. A good communicator must be a good listener. Become used to listening to people's stories and read the biographies of the scientists you admire. That exploration might teach you that there is a human being behind each scientist. On the History Channel, you might watch interesting stories about people who made striking contributions to our history: http://www.history.com/speeches.

Also, there are countless websites and online videos regarding approaches to improve your communication skills; see, for example, 68 funny videos on: http://www.howcast.com/guides/520-How-to-Improve-Your-Communication-Skills.

Chapter 11

Supporting Materials

U.P. Albuquerque, *Speaking in Public About Science*,
DOI 10.1007/978-3-319-06517-5_11,
© Springer International Publishing Switzerland 2015

The texts below are intended for you to perform a critical reflection as well as to train the tasks suggested in this book. These texts are translations from chronicles that I have published in our blog.

Text 1: I Am a Doctor, and You?

Once, I was invited to teach a course on my field of expertise in another country. In addition to the challenge posed by the need to express my ideas in another language, I had to confront a situation that was curious at least. After teaching the two first days of the course, I confided to a colleague, "*I believe the students do not like my classes. They remain totally mute and as if frozen.*" My colleague answered, "*This is how it is here!*" Indeed, later on a student explained to me, "*Professor, here, students should not express themselves in the classroom. Once, I asked a professor what books should I read for his class, and he answered: 'None! You just have to pay attention to what I say!'*"

That discussion cleared away my doubts and also revealed a particularity, to wit, the principle or *argument from authority*. The argument from authority is a resource many among us use to strengthen our arguments, even when we are not at all sure about them. In science, the argument from authority might be used in two ways. First, as Bruno Latour explains in *Science in Action*, one might look for "prestigious" or "authoritative" allies to impress an interlocutor or ideological opponent. This facet of the argument from authority might make someone accept some ideas even when the ideas are poorly substantiated as a function of the speaker's "status." If that acceptance might happen in the academic milieu, imagine in other settings! Once, I was at the bank cashier line, and a woman started talking very enthusiastically about a TV interview of a famous doctor. Her interlocutor said that she believed a particular assertion to be groundless, to which the first speaker, quite distraught and almost screaming, replied, "*Do you pretend to know more than Dr. X?*"

I do not mean to say that to use references to strengthen an argument is wrong, but I mean that some ways in which they are used are. In that same book, Bruno Latour observes, "*an essay without references is like a helpless child in a large and unknown city at night: it is alone, lost, and anything might happen to it.*" So I looked in Latour's methodological rules for a strategy of protection against the "negative" side of the argument from authority, and I found the following statements:

1. Observe the setting in which the allegation or arguments chosen for analysis are placed.
2. Identify the individuals who are collecting the data to shift the "status" of allegation to that of fact, and identify the individuals who are doing the exact opposite.
3. Observe the trajectory of that allegation as a function of the clash between the two parties mentioned above.

Readers might find it difficult to perform all those steps whenever they are confronted with a new argument or allegation. However, that is precisely the type of reasoning that characterizes science.

The second negative facet of the argument from authority is, perhaps, more difficult to counterargue. It might be called *argument from head-professorship* (I have just made it up!) and can be illustrated by the story at the beginning of this text, which can be summarized in one sentence: *I am THE professor!* Once, an unhappy PhD candidate came to me at a conference and said, "*I tried to debate some of the lecturer's ideas, but he ended the discussion saying that the only one with a PhD there was him!*" A lecture or written work without sound foundations might denote the weakness of the ideas themselves, and thus, the reaction of the lecturer might have been due to lack of self-confidence. Lack of self-confidence often elicits a self-defense strategy based on attacks or protection. One such strategy is to build a wall of "authority" to avoid discussions or debates. In turn, lack of self-confidence might arise from a negative self-image resulting from a given worldview that permeates society (e.g., unfair competition) and that has a natural home in the academic milieu.

The following text published by Roberto DaMatta in the newspaper O Globo (October 24, 2007), and quoted by Miriam Goldberg in her book, Noites de Insônia (*Sleepless Nights*), might be useful to stimulate our reflection. Before complaining that this text has nothing to do with the arguments I have discussed up to this point, please, read to the end.

> Envy is a basic feeling in Brazil. A non-envious Brazilian has yet to be born. This is so powerful, that we use the verb "to have" rather than "to feel" when we speak about envy. Other human beings and other peoples feel envy (i.e., one of many feeling), but not us; envy possesses us. Taken by the perverse and human conjunction of hatred and displeasure, precisely elicited by someone else's success. Our problem is [...] that a bright individual who, we believe, "takes" (steals, erases, represses, hinders) from us the chance to shine in that region beyond the sky. Envy, for me, is the strongest sign of a closed system in which individual autonomy is weak, and everyone spends his or her lives

comparing oneself to another. Control by means of intrigue, rumor, gossip, tittle-tattle, is a proof of that incessant comparison of behaviors, which does not aim at making all equal, but to establish hierarchies, to distinguish, to grade. The horror of competition, common sense, transparency, and mobility is the other side of that culture in which to be successful is illegitimate, a disaster, and even a crime. How, then, will one not be envious of someone else's success, being as we are persuaded that success is an act of treason against the compliant and obedient collective to which we belong? How, then, one will not feel envious, when successful people are the ones who refuse to behave as a sheep that does not attract attention, to rather become the most showy one— i.e., the symbol of selfishness and ambition?

Some people give way to the pressure, and in their despaired attempt to reconcile their personal demands with the external expectations, in that emotional, fully non-rational collision, the best argument they can adduce is **I am a Doctor, and you?**

Text 2: What Is a Thesis?

I entered my scientific career at a time when researchers used to write mega-theses. Let me explain. The master's dissertations and doctoral theses were huge, and thus, my academic immaturity led me to believe that the larger the number of pages in a dissertation, the greater the impact it would make (well, as a matter of fact, that is also true, but that impact is not necessarily a scientific one!).

In time, a new perspective developed, and the traditional dissertation or thesis was replaced by a more compact version. Many graduate programs allowed candidates to write independent chapters, each one corresponding to an article to be submitted to some specialized journal. The reasons behind that shift were many: to make sure that academic essays would be effectively published; to improve the scientific production of graduate programs and professors; to meet the demands of the governmental evaluation agency (CAPES) to improve the quality of graduate programs; and also—although no one dares to admit it—to facilitate the task of the dissertation committees.

Some people still believe that the traditional model for dissertations and theses was the ideal one because it enables researchers to report on the full experience he or she underwent while trying to answer the scientific questions he or she posed. In contrast, other people adduce that published articles make studies spread faster, thus avoiding, as Mário Prata would say, the condemnation of studies to become historical footnotes stored in dusty libraries. Both sides put forth good arguments in support of their views, but I believe that the problem here is that both of these positions are extreme.

I believe that much of the information included in dissertations and theses might be discarded because it does not contribute to make reading more pleasant and would also not be published. The dissertations with article layout are at risk of becoming excessively objective and synthetic, when some particular features are not carefully taken into consideration. However, two crucial questions should be answered before addressing that polarized debate, to wit: (1) What is a thesis? (2) What type of professionals do we want to train?

What Is a Thesis?

There is much to do and to learn about nature, life, and people. A scientist is an individual eager to always learn a little bit more who looks for new methods to answer questions that trouble him or her. And once the answers are found, he or she will pose new questions. It is a fantastic and fascinating vicious circle. Dissertations are a product of a part of a researcher's life, in which he or she records that never-ending experience of asking and answering. For some, developing a thesis is pleasurable, while for others it is painful. In *Noites de Insônia* (*Sleepless Nights*), the anthropologist Miriam Goldberg comments that the sociologist Howard Becker "did not suffer much" to complete his training, while she is

> ...one of that type of people who suffer all the time: when writing a course final essay, preparing and defending the dissertation, to obtain a postdoctoral grant, to apply for a professorship at the Institute of Philosophy and Social Science (Instituto de Filosofia e Ciências Sociais—IFCS). And when I believed I could finally relax, endless suffering began: when requesting a grant to the National Council of Scientific and Technological Development (Conselho Nacional de Desenvolvimento Científico e Tecnológico—CNPq), when renewing the CNPq grant, when writing articles to be published in prestigious scientific journals, when writing books and expecting that people would read them and like them, when supervising students, when facing the criticism made by dissertation committees to my supervisees, when presenting papers at the National Association of Graduate Studies and Research in Social Science (Associação Nacional de Pós-Graduação e Pesquisa em Ciências Sociais—ANPOCS) and the Brazilian Association of Anthropology (Associação Brasileira de Antropologia—ABA), when giving lectures, when teaching undergraduate and graduate programs, and so forth.

I prefer to believe that such is a quite natural type of suffering, and as a matter of fact, I like it very much because it is a part of the scientist's job description. I do not allude to suffering as such but to the experience it affords. A thesis allows for the full unfolding of a given experience because it is the place where the many elements constructed and discovered in the course of years articulate to make sense. A thesis serves to bring some light to a place up to then only occupied by the darkness of doubt and uncertainty, or even to generate more doubts and uncertainties. A thesis is an integral part of a researcher's experience, and as such, it should carry its mark and reflect the skills required from a scientist-in-training: to know how to think, to know how to pose questions. The genuine work begins only the day after the dissertation defense. That is the moment when the dust starts settling, and scientific maturity shows the first hesitating steps of its journey. Nothing can be more natural in any profession!

What Type of Professionals Do We Want to Train?

In his book *Diálogos com Cientistas* (*Dialogues with Scientists*), Professor Eloi Garcia paraphrases Einstein to emphasize that posing questions is the most essential aspect. More than posing questions, it is necessary to ask the right questions. According to Professor Garcia, a scientist must do science in a humane manner. That is, he or she should be productive, open paths, draft new scenarios, and be flexible when thinking. A little more pragmatically but without dismissing the values described in Garcia's book, I believe that scientists should be able to think, to reach beyond the mere reproduction of laboratory protocols or experiments designed by someone else. Scientists ought to be able to make criticisms and self-criticism on a steady basis. I do not allude to paralyzing criticism but to criticism that is constructive and helps to set new directions. In addition, a scientist-in-training should be taught about the freedom of thought and that type of "guided" autonomy that leads to self-confidence.

Returning to the format of scientific writing, none of the models discussed above ensure that the abovementioned qualities and skills will be incorporated and developed. I have read theses in traditional format, whose authors were unable to defend their content, as well as articles so synthetic and signed by so many authors, that it was impossible to identify the intellectual contribution of the master's or PhD student. I believe that the best product we might hope for is a highly qualified and trained professional. Once this is achieved, we will certainly produce good scientific articles and a new type of scientist also able to train qualified human resources. For that purpose, I believe that a slight reformulation of the graduate programs is in order. For instance, I have systematically pointed to the need for scientists-in-training to have opportunities to develop their writing and speaking skills. In addition, philosophy of science ought to be mandatorily included in the curriculum. How might one expect to train scientists without showing to them what science is?

It seems that the problem is the same as the one we have to address in the public calls for tenured positions: lack of preparation. In her book, *Nos Bastidores da Ciência* (*The Backstage of Science*), Professor Debora Menezes emphatically corroborates that observation. We attend graduate programs, we are prepared for conducting research, but we are not prepared to teach. Nevertheless, in the calls for tenured positions and later on in our professorial careers, we are also assessed regarding our teaching skills.

What Do We Want for the Future?

We must think our policies over. What is more relevant for the advancement of science and technology in Brazil: one article published in a high-impact journal, or a high-impact scientist? Please, do not answer that both of these things are inherently interconnected because they are not! Therefore, the format of a thesis matters very little when there is a well-trained professional behind it, one who is able to take ownership of that which is written.

Text 3: The Science I Imagine and the Science There Is

Once, I entered a hot debate with a friend. The topic of the discussion was science. He argued that science is objective, neutral, and unbiased by definition. I agreed only with the expression "by definition" because science is very seldom objective, neutral, and unbiased in reality. I am sure that assertion will be read with dismay by purists, i.e., the ones who viscerally defend those "myths" about science. Science is a product of the human intellect, made by people and for people. Like every human product, science is impregnated with subjectivity and bias. This condition is how human beings are. The truth is that we, as scientists, try to be objective, neutral, and unbiased. However, that is not possible most of the time.

I remember an assertion made by Lynn Margulis about her work regarding the "Endosymbiotic Theory." Lynn Margulis was born in 1938 and earned a PhD in UC Berkeley in 1963. Her theory caused some scientific chaos because it ran against the ideas prevalent at that time as well as against the *status quo* in science, which was then controlled by male scientists. While observing the cell organelles, scientists found that mitochondria and chloroplasts, e.g., have their own DNA. What does this finding mean? I used to tell my students that it is as if we have small "aliens" inside us because the genetic material of those organelles is completely different from the material in the remainder of the cells of an animal or plant body. Margulis suggested that prokaryote cells were "taken inside" by eukaryote cells in the course of cell evolution. Thus, the "incorporated" beings entered a type of symbiotic life with the eukaryote cells. I always found that discovery fascinating because in addition to its logical implications for the understanding of cell biology, it made Margulies argue later that as concerns evolution, "cooperation" is as meaningful as "competition" (which is a notion that dominates in evolutionary biology). Margulis' paper was rejected several times (approximately 15!) before it was published in *Journal of Theoretical Biology*. Publication was followed by criticism and critics (these are two different things, dear reader, because, fortunately or unfortunately, some people are specialists in merely criticizing). However, Margulis won the war! Margulis' idea had been initially rejected for several reasons: (1) Because it ran against the prevailing paradigm. (2) Because a woman was the one who ran against the prevailing paradigm. (3) Because she dared to do science at a time when science was ruled by a dominant male group.

I told my friend that story, and he jumped to answer, "*That doesn't happen anymore! Our society advanced as concerns the ideals of equality! Women earned their place, especially in science.*" I cannot disagree with that statement; however, we are still quite far from the ideal situation. In fact, many things are still at the level of ideas. Some studies of scientometrics found evidence indicating that women's papers are less cited than papers authored by males in some scientific fields. There is also evidence showing that Latin American authors are less cited than their North American and European peers, even when all of them publish in the same journals. I do not believe that the problem is that women or Latin Americans produce low-quality science. Low-quality science has no color, religion, race, or citizenship, and the same applies to high-quality science. However, we cannot deny the fact that in some places, scientists have better conditions and admirable infrastructure to produce science, although such fact is no warranty for products of superior quality.

The human side of scientists also comes to the foreground when their beliefs are challenged, their hypotheses are questioned, or their pride is hurt. The biologist Edward Wilson, an expert on social insects and doubtlessly one of the brightest minds in modern biology, elicited strong emotional reactions upon founding *sociobiology*. According to this field of science, much of human behavior might be explained based on evolution and genes. A more radical version of sociobiology sought to fully eliminate the role of culture. Wilson's first writings were quite radical, and over time, he developed a softer version of sociobiology. However, in his initial phase, he was the target of heavy criticism and eventually the victim of physical attack. The verbal attacks were countless. One of his main advocates was Richard Dawkins, who is perhaps the scientist who contributed most to the dissemination of the hypothesis, positing a genetic or biological basis for human behavior with his book "The Selfish Gene." Here, I am not taking any position regarding that subject, I am merely seeking to illustrate how passionate, and even irrational, scientists might be when someone appears and challenges the prevailing ideas. Let Wilson tell us a striking episode in his scientific career:

> On August 1, 1977, sociobiology was on the cover of Time. On November 22, I received the National Medal of Science from President Carter for my contributions to the new discipline. Two months later, at the annual meeting of the American Association for the Advancement of Science, held in Washington, demonstrators seized the stage as I was about to give a lecture, dumped a pitcher of ice water on my head, and chanted, "Wilson, you're all wet." The ice-water episode may be the only occasion in recent American history in which a scientist was physically attacked, however mildly, simply for the expression of an idea. How could an entomologist with a penchant for solitude provoke a tumult of this proportion?[1]

I believe that we will advance toward satisfactory scientific education when scientists-in-training are confronted with the actual conditions of science production and when they are challenged to ponder on the beliefs and ideologies that are presented to us.

---

[1] Translated from the Portuguese edition of the Book: *Naturalista* (Naturalist) of Edward Wilson. Rio de Janeiro: Nova Fronteira, 1997.

References

Barrass R (1991) Os cientistas precisam escrever [Scientist must write]. T.A. Queiroz Editor, São Paulo

Barrass R (2002) Scientist must write: a guide to better writing for scientists, engineers and students. Routledge, New York, NY

Briscoe MH (1995) Preparing scientific illustrations: a guide to better posters, presentations, and publications. Springer, California

Camara JM Jr (1993) Manual de expressão oral & escrita [Manual of oral and written expression]. Vozes, Rio de Janeiro

Carnegie D (1994) Como falar em público e influenciar pessoas no mundo dos negócios [The quick and easy way to effective speaking]. Record, Rio de Janeiro

Jacob F (1988) A estátua interior [The statue within: an autobiography]. Dom Quixote, Lisboa

Krausz RR (1986) Administre bem o seu tempo [Manage your time]. Nobel, São Paulo

Lucas SE (2003) A arte de falar em público [The art of public speaking]. Livros Técnicos e Científicos Editora S.A, Rio de Janeiro

Medawar PB (1982) Conselho a um jovem cientista [Advice to a young scientist]. Editora Universidade de Brasília, Brasília

Morgan S, Whitener B (2006) Speaking about science: a manual for creating clear presentations. Cambridge University Press, Cambridge

Silva MALS (1991) Iniciação à comunicação oral e escrita [Introduction to oral and written communication]. Editorial Presença, Lisboa

Vanoye F (1993) Usos da linguagem [Uses of language]. Martins Fontes, São Paulo

U.P. Albuquerque, *Speaking in Public About Science*,
DOI 10.1007/978-3-319-06517-5,
© Springer International Publishing Switzerland 2015

Index

A

Argument from authority
    argument from head-professorship, 57
    Latour's methodological rules, 56
    self-defense strategy, 57

B

Blackboards, 45–46
The body speaks, 27
Bulletin boards/posters, 17–18

C

Communication skills
    expository practices, 8–9
    psychological and emotional factor, 9
Conferences and lectures, 16
Criticisms, 60

D

Dialogues with Scientists, 60
Dissertation defenses, 60
    awkward situations, 19
    definition, 17
    mistakes, rectification of, 20
    presentation rehearsal, 18
    speaker's posture and vocabulary, 19
    speaking fluently, 20

E

Endosymbiotic theory, 61

L

Lasswell's model, 52

M

Margulis, Lynn, 61–62

N

Nervousness, 27, 28

O

Ockham's razor, 34
Online resources, 53

Oral communication, 2
    efficacy and fruitfulness, 12
    features and obstacles, 14
    and personal relationships, 12
    professionals experiences, 12
    in scientific discourse, 12
    scientific presentations, 17
    self-knowledge, 13
    social coexistence, 12
Oral expression, 3
Oral presentations, 2

P

Philosophy of science, 60–61
Posters, 17–18
Practical knowledge, 3
Psychological crutches, 27

R

Roundtables, 16–17

S

Science
    definition, 61
    sociobiology, 62
Scientific arrogance, 24
Scientific discourse
    accuracy, 33–35
    clarity, 32–33
    completeness, 33
    comprehensive study, 38
    conclusion, 40, 41
    consistency/unity, 35
    definition, 38
    discussion, 40
    excitement and enthusiasm, 35
    explanation, 31
    impartiality, 33
    introduction, 40
    methodology, 40
    objectivity, 33–34
    persuasion, 36
    results section, 40

U.P. Albuquerque, *Speaking in Public About Science*,
DOI 10.1007/978-3-319-06517-5,
© Springer International Publishing Switzerland 2015

Scientific discourse (*continued*)
  script preparation, 41–42
  sentence order, 33
  simplicity, 34
  sincerity, 35–36
  subject delimitation, 38
  subject exploration, 38
  subject mastery and reflection, 38
  types of, 41
Scientific modesty, 24, 36
Scientific presentations
  audience, empathy, 25
  bulletin boards/posters, 17–18
  conferences and lectures, 16
  confidence, 26
  dissertation defenses
    awkward situations, 19
    definition, 17
    mistakes, rectification of, 20
    presentation rehearsal, 18
    speaker's posture and vocabulary, 19
    speaking fluently, 20
  expression features, 20
  goal of, 25
  language vices, 27
  monotonous speakers, 26
  naturalness, friendliness,
      and energy, 26
  nervousness, 27, 28
  oral communications, 17, 26
  poster sessions, 26
  predisposing factors, 25
  presentation styles, 26
  psychological features, 21
  read speeches, 29
  roundtables, 16–17
  scientific arrogance, 24
  self-knowledge and self-improvement, 26
  seminars, 16
  speaker's posture, 27, 28
  technical features, 20
  technical term, 29
  time availability, 30
    vocabulary, 29
    voice inflection and articulation, 27
Scientific writing, 32
Scientists, 48
Self-criticism, 60
Self-knowledge, 13
Seminars, 16
Sleepless Nights, 57, 59
Slides, 44–45
Social coexistence, 12
Sociobiology, 62
Speeches
  audiences, 51
  conclusion, 51
  discourse plans and points of view, 50
  improvement, 52
  informational discourse, 51
  Lasswell's model, 52
  organization training and
      preparation, 50
  persuasive speech, 51
  preparation and development, 51
  scientific model, 52
  speaker's positive qualities and flaws, 51
  story reconstruction, 52
  visual resource preparation, 52

T
Theoretical knowledge, 3
Thesis
  doubts and uncertainties, 60
  researcher's experience, 60
  traditional dissertation, 58

V
Visual resources
  blackboards, 45–46
  rules, 44
  slides, 44–45
Vocabulary, 3, 29

W
Wilson, Edward, 62–63